T0247681

Managing Military Personnel Costs

Operation Retrenchment Specter, A Workforce
Futures Game

MATTHEW WALSH, LISA M. HARRINGTON, THOMAS LIGHT

Prepared for the Department of the Air Force
Approved for public release; distribution unlimited

 PROJECT AIR FORCE

For more information on this publication, visit **www.rand.org/t/RRA1218-2**.

About RAND

The RAND Corporation is a research organization that develops solutions to public policy challenges to help make communities throughout the world safer and more secure, healthier and more prosperous. RAND is nonprofit, nonpartisan, and committed to the public interest. To learn more about RAND, visit www.rand.org.

Research Integrity

Our mission to help improve policy and decisionmaking through research and analysis is enabled through our core values of quality and objectivity and our unwavering commitment to the highest level of integrity and ethical behavior. To help ensure our research and analysis are rigorous, objective, and nonpartisan, we subject our research publications to a robust and exacting quality-assurance process; avoid both the appearance and reality of financial and other conflicts of interest through staff training, project screening, and a policy of mandatory disclosure; and pursue transparency in our research engagements through our commitment to the open publication of our research findings and recommendations, disclosure of the source of funding of published research, and policies to ensure intellectual independence. For more information, visit www.rand.org/about/research-integrity.

RAND's publications do not necessarily reflect the opinions of its research clients and sponsors.

Published by the RAND Corporation, Santa Monica, Calif.
© 2023 RAND Corporation
RAND® is a registered trademark.

Library of Congress Cataloging-in-Publication Data is available for this publication.
ISBN: 978-1-9774-1140-2

Cover: U.S. Air Force photo by Sean M. Worrell.

Limited Print and Electronic Distribution Rights

About This Report

The military personnel (MILPERS) budget provides financial resources to compensate active-duty personnel. This includes pay and allowances, health care and retirement pay accruals, permanent change of station (PCS) travel, and other MILPERS costs. Spending on MILPERS has grown at an average annual rate of 3.3 percent since fiscal year (FY) 2000 to approximately $36 billion in FY 2021. Along with the growth of the MILPERS budget, the average cost of an airman has increased. After normalizing for the size of the workforce, the average cost of an airman increased by 106 percent from FY 2000 to FY 2021. By comparison, civilian pay grew by only 60 percent during the same period. The increased cost of personnel threatens to undermine the Department of the Air Force's (DAF's) ability to field a ready workforce, and the resulting growth in the MILPERS budget threatens to divert resources from modernization and sustainment efforts.

To limit the growth of the MILPERS budget, the DAF must systematically evaluate a variety of policy options and then select policies that control personnel costs without increasing risks to mission and force. This requires an enterprise-wide perspective. To advance such a perspective, RAND Project AIR FORCE (PAF) conducted a workforce futures policy game—Operation Retrenchment Specter—with the U.S. Air Force's (USAF's) most senior human resources leader, other senior leaders, and more than 50 USAF participants. Teams competed to generate novel options to limited MILPERS costs without incurring unacceptable risks. To ensure that the game construct reflected real-world trade-offs accurately, teams used a modeling ecosystem to simulate the monetary and nonmonetary effects of workforce and personnel policy changes. This report describes the game and the options that teams generated.

The research reported here was commissioned by the Director of Manpower, Organization and Resources, Headquarters U.S. Air Force (AF/A1M) and conducted within the Workforce, Development, and Health Program of PAF as part of an FY 2021 project "Manpower/Personnel Realignment Tool." A companion report titled *Assessing the Implications of Policy Options for the Military Personnel Budget: An Analytic Framework for Evaluating Costs and Trade-Offs* provides additional context and supporting analyses.

RAND Project AIR FORCE

RAND Project AIR FORCE (PAF), a division of the RAND Corporation, is the Department of the Air Force's (DAF's) federally funded research and development center for studies and analyses, supporting both the United States Air Force and the United States Space Force. PAF provides the DAF with independent analyses of policy alternatives affecting the development, employment, combat readiness, and support of current and future air, space, and cyber forces.

Research is conducted in four programs: Strategy and Doctrine; Force Modernization and Employment; Resource Management; and Workforce, Development, and Health. The research reported here was prepared under contract FA7014-16-D-1000.

Additional information about PAF is available on our website: www.rand.org/paf/

This report documents work originally shared with the DAF on March 13, 2022. The draft report, dated March 2022, was reviewed by formal peer reviewers and DAF subject-matter experts.

Acknowledgments

We thank Lt Gen Brian Kelly, deputy chief of staff for Manpower, Personnel, and Services (AF AF/A1), for supporting the workforce futures game, and Brig Gen Gentry W. Boswell, director of Manpower, Organization and Resources, deputy chief of staff for Manpower, Personnel and Services, Headquarters U.S. Air Force (AF/A1M), for sponsoring the project. We also thank Col Patrick White (AF/A1M) and Col James Barger (AF/A1M) for their help and expertise while developing the workforce futures game.

We thank the senior leaders who provided final adjudication during the workforce futures game: Joseph McDade, assistant deputy chief of staff for Plans and Programs (AF/A8); Lt Gen Samuel Hinote, deputy chief of staff for Strategy, Integration and Requirements (AF/A5); Lt Gen Warren Berry, deputy chief of staff for Logistics, Engineering and Force Protection (AF/A4); Maj Gen Charles Corcoran, assistant deputy chief of staff for Operations (AF/A3); and Brig Gen Frank Verdugo, director of Budget Operations and Personnel (SAF/FMBO). We also thank the subject-matter experts from the Air Force for their participation, especially those from ACC/A9, ACC/TES, AFIT, AFMAA, AWC, AF/A1C, AF/A1D, AF/A1H, AF/A1I, AF/A1M, AF/A1P, AF/A1X, AF/A3, AF/A4, AF/A5, AF/A8, AF/A8P, AF/A8X, SAF/FMB, SAF/MRM, USAFA/OLEA, and USSF/S1.

Finally, we thank the many RAND colleagues who helped with this work—principally, but not exclusively, Clara Aranibar, John Ausink, Ray Conley, Sarah Denton, Christopher Ferris, Erin Leidy, Nelson Lim, Sarah Lovell, Miriam Matthews, Libby May, Al Robbert, Anthony Rosello, and Julia Vidal Verastegui.

Summary

Issue

The military personnel (MILPERS) budget provides financial resources to compensate active-duty personnel. Spending on MILPERS has grown at an average annual rate of 3.3 percent per year since fiscal year (FY) 2000, to approximately $36 billion in FY 2021. To ensure a ready workforce without undercutting modernization and sustainment efforts, the Department of the Air Force (DAF) must explore options to limit MILPERS costs. At the same time, DAF must consider the nonmonetary trade-offs and risks that these options entail.

This is a *wicked problem*:[1] It lacks clear boundaries because any action may have repercussive effects throughout the DAF enterprise, and it lacks one "correct" solution because the DAF stakeholders prioritize different financial and operational objectives over varying time horizons. To tackle this problem, the DAF must engage diverse stakeholders to create a strong information-sharing environment, generate creative options, and build consensus around those options.

Approach

RAND Project AIR FORCE (PAF) designed and conducted a workforce futures policy game—Operation Retrenchment Specter—with support from the U.S. Air Force's (USAF's) most senior human resources leader. During the game, teams competed to find options to limit MILPERS costs without introducing unacceptable risks. To ensure the game construct reflected real world trade-offs accurately, teams used a modeling ecosystem to simulate the monetary and nonmonetary effects of workforce and personnel policies in real time. In addition to proposing options, teams developed hedges and other shaping actions to mitigate risk. At the conclusion of the game, teams presented options to senior leaders from the Air Staff and the Air Force Secretariat. Options yielded projected annual savings of $500 million to $2 billion.

Key Findings

Teams proposed a diverse set of options (Table S.1). Two involved reducing manpower requirements by consolidating or eliminating organizations or functional communities (Gold team and Black team); two involved shifting to a more junior grade mix (Silver team and Blue

---

[1] *Wicked problems* are a class of planning and policy problems that are difficult to tackle because they lack clear definitions and boundaries, they involve complex interdependencies, and they do not have "correct" solutions. Horst W. J. Rittle and Melvin M. Webber, "Dilemmas in a General Theory of Planning," *Policy Sciences,* Vol. 4, No. 2, 1973.

team); and one involved converting officer positions to the enlisted force (Green team). These options revealed a fundamental trade space among cost, size of the workforce, and experience.

Table S.1. Comparison of Outcomes from the Baseline Scenario and Proposed Solutions

Option	MILPERS Cost (billions of dollars)	Size (full-time equivalents)	Experience (years of service)
Baseline	35.0	331,533	7.5
Gold: Flatten organizational structure to enable reductions in end strength.	32.8	312,292	7.5
Black: Consolidate Air Force Specialty Codes to enable reductions in end strength.	34.3	324,902	7.5
Silver: Limit annual pay increases and shift to a more junior grade mix.	34.6	331,533	7.3
Blue: Reduce administrative overhead to shift to a more junior grade mix.	34.5	331,533	7.4
Green: Convert officer positions to the enlisted force.	34.7	331,533	7.5

NOTE: Cells in dark red and dark green denote a 2.0 percent or more change from baseline. Cells in light red and light green denote a 1.0 percent or more change from baseline.

As seen during the game, senior leaders must decide which compromises to make as they generate future workforce designs.

- To trade off size, the Air Force could consolidate or eliminate organizations, installations, or functional communities.
- To trade off experience, the Air Force could more fully utilize the talent and abilities of junior service members and enlisted personnel.
- To reduce MILPERS costs without *directly* trading off size or experience, the Air Force, with congressional approval, could limit growth in basic pay. However, this could drive recruiting or retention trends that indirectly trade off size and experience.

These options entail implementation risk.

- Options requiring congressional approval (e.g., pay) have high implementation risk.
- The Air Force already has the authority to implement many of the options considered here (e.g., alter promotion timing, alter special and incentive pay, reduce grade ceilings, and consolidate organizational structures and functional communities). However, the impetus to do so has not been strong enough to overcome inertia.

The DAF could take hedging and shaping actions to reduce risk.

- For options that reduce size, the Air Force could adopt technologies to increase workforce efficiencies, or it could reduce the number of days that service members spend in student or transient status to accrue manpower savings.
- For options that shift to a more junior grade mix, the Air Force could delay promotions to give individuals more time to develop.
- For options that reduce retention in the active-duty force, the Air Force could leverage programs to encourage service members to enter the reserves to retain their experience.
- For options that reduce basic pay, the Air Force could use special and incentive pay in a more targeted manner to sustain retention in high-demand career fields or in those with high production costs.

The savings from most options, although modest in relative terms, are significant.

- Because the MILPERS budget is so large, a 2 percent savings could be repurposed to support thousands of additional personnel or tens of thousands of flying hours.

Recommendations

- Because the options shown in Table S.1 have sweeping implications, the Air Force could establish a group that reports to the Air Force's chief of staff that is dedicated to developing, vetting, and prioritizing actions to limit MILPERS costs.
- Given time constraints, teams effectively proposed solution classes rather than formal options. The benefit of the game was determining which solution classes are promising enough to warrant further attention. The Air Force should return to data-driven and other evidence-based approaches to optimize solutions within these classes.
- The Air Force should continue to use policy games to explore ways to control MILPERS costs. In particular, the Air Force could develop scenarios involving different economic conditions and/or contingencies to determine how various solutions fare. In addition, the Air Force could include players from different communities and organizations in future games to enable generation of new ideas and to establish more-widespread buy-in.
- Given that domestic and international contexts are not static, the Air Force should periodically revisit workforce design and exercise options to rebalance the force.
- Most workforce and personnel planning problems that the Air Force faces are wicked problems. The Air Force should include policy games in the set of evidence-based methods routinely used to examine such problems.

Contents

Figures and Tables

Figures

Tables

Chapter 1. Introduction

The military personnel (MILPERS) budget provides financial resources to compensate active-duty personnel. This includes pay and allowances, health care and retirement pay accruals (RPAs), permanent change of station (PCS) travel, and other MILPERS costs. As the Department of the Air Force (DAF) plans and programs its budget, including the MILPERS budget, it must field the workforce needed to deliver capacity and capability to combatant commanders today while ensuring financial flexibility for operations and future needs. As the DAF noted in its MILPERS program budget documentation for fiscal year (FY) 2022, "Our biggest leadership challenge is taking care of people while striking the right balance between maintaining today's readiness and posturing future modernization and recapitalization priorities."[1]

In FY 2021, U.S. Air Force (USAF) end strength equaled nearly 335,000 active component officers, cadets, and enlisted personnel.[2] The MILPERS budget provides the pay and allowances for these individuals. At a cost of approximately $36 billion, the MILPERS budget made up more than 20 percent of the DAF's total FY 2021 budget of $168 billion.[3] For comparison, only the Operation and Maintenance (O&M) budget was larger, at $60.9 billion.[4] Between FY 2000 and FY 2021, spending on active-duty personnel grew at an average annual rate of 3.3 percent, outpacing growth in prices in the overall economy, which averaged 1.9 percent per year for the same period.[5] At the same time, the average cost of an airman has risen. After normalizing for the size of the workforce, the average cost of an airman increased 106 percent from $50,000 in FY 2000 to $103,000 in FY 2021. By comparison, civilian pay grew by only 60 percent during the same period.[6]

---

[1] DAF, *Fiscal Year (FY) 2022 Budget Estimates: Military Personnel Appropriations*, May 2021a.

[2] FY 2021 end strength (as of September 30, 2021) equaled 64,936 officers, 266,451 enlisted personnel, and 4,098 cadets (335,485 total) at a cost of $35,862,533,000. Average work years for FY 2021 were slightly higher, at 349,460; see DAF, 2021a.

[3] The FY 2021 active-duty MILPERS budget totaled $35,862,533,000 and made up about 21 percent of the DAF's total budget of $168,237,000; see DAF, 2021a.

[4] At $26.1 and $26.6 billion, respectively, the DAF's budgets for procurement and research, development, test, and evaluation were less than the MILPERS budget.

[5] The FY 2000 active-duty MILPERS budget totaled $17,978,193,000; see DAF, *FY 2002 Amended Budget Submissions to Congress June 2001: Operation and Maintenance, Air Force*, Vol. 1, June 2001. Growth in prices in the overall economy between FY 2000 and FY 2021 is measured using the gross domestic product price deflator; see Office of the Under Secretary of Defense (Comptroller), *National Defense Budget Estimates for FY 2022*, (Green Book), August 2021.

[6] Rates of civilian pay growth were calculated by the RAND Corporation using Table 5.1 and Table 5.5 of Office of the Under Secretary of Defense (Comptroller), 2021.

The growth in the average cost of an airman threatens to undermine personnel readiness, while the resulting growth in the MILPERS budget threatens to divert resources from vital military modernization efforts. The chief of staff of the Air Force (CSAF) has stated unequivocally:

> No matter what happens with the budget, it will require us to make tough choices. We need to continue developing a lethal and affordable force that Congress supports. Action Order D drives the Air Force to "make force structure decisions . . . and amend force planning processes to create the fiscal flexibility required to design and field the future force we need."[7]

This is a crippling problem. The workforce is the foundation for readiness, yet the DAF, including the U.S. Air Force (USAF), must maintain fiscal flexibility to meet other immediate and future needs. This is a complex problem. Many options could potentially limit growth in the MILPERS budget, yet those options may introduce unacceptable risks throughout the USAF enterprise. This is a problem without a "correct" solution. Different stakeholders may evaluate options based on different financial and operational trade-offs over varying time horizons. Finally, this is an enduring problem. As domestic and international contexts change, the USAF will need to revisit workforce design and exercise options to rebalance the workforce.

Decisions about workforce structure are data driven in the sense that they use historical data, econometric models, staffing models, and cost and planning factors. In practice, data-driven approaches to policymaking entail certain key assumptions: (1) historical patterns are generalizable to new policies under consideration; (2) current conditions will hold in the future; (3) data are sufficiently rich to establish causal models linking policies to outcomes; and (4) the costs and risks can be quantified and estimated using a data-driven approach.[8] The first three assumptions involve the feasibility of using data-driven approaches to project outcomes, and the fourth involves the feasibility of using data-driven approaches to capture substantive costs and risks.

These assumptions are not fully met in the case of evaluating policy options to reduce MILPERS costs. Because of the interrelated nature of manpower, personnel, and services functions—not to mention their relationship with combat operations—it is not feasible to create a causal model that encompasses the complete problem. In addition, the problem entails significant unknowns, such as future economic conditions that may influence recruiting and retention, and operational demands that may drive personnel needs. Finally, the problem encompasses a wide variety of outcomes that play out across different timescales. Not all of these can be quantified, and policy choices depend on subjective judgments about the importance of each.

---

[7] Charles Q. Brown Jr., *CSAF Action Orders to Accelerate Change Across Air the Force*, Chief of Staff, U.S. Air Force, February 7, 2022.

[8] Elizabeth M. Bartels, Jeffrey A. Drezner, and Joel B. Predd, *Building a Broader Evidence Base for Defense Acquisition Policymaking*, RAND Corporation, RR-A202-1, 2020.

For these reasons, the problem of limiting growth of the MILPERS budget cannot be solved by using standard analytic methods within the silo of manpower, personnel, and services. To tackle this problem, the USAF must engage diverse stakeholders to create a strong information-sharing environment, generate creative options, and build consensus around those options.

This report describes a workforce futures policy game—Operation Retrenchment Specter—designed to bring an enterprise-wide view to the challenge of limiting MILPERS costs. The remainder of this report is organized as follows: Chapter 2 provides background on MILPERS spending and outlines the challenges of finding viable options to limit its growth. In Chapter 3, we describe the design of the workforce futures policy game. Chapters 4 and 5 present the results of the game. In Chapter 6, we summarize the findings and offer recommendations.

Chapter 2. The Rising Cost of Military Personnel

In this chapter, we describe elements that contribute to standard composite pay rates—the main determinant of the average cost of an airman—before presenting a historical analysis of MILPERS spending. We then situate MILPERS costs within a broader dynamic system that includes workforce and personnel policies to show how these may affect MILPERS spending.

Elements of Standard Composite Pay Rates

Air Force MILPERS spending can be estimated from (1) the size and grade mix of the workforce and (2) the standard composite pay rates reflecting the cost of an airman. The standard composite pay rates published by the Deputy Assistant Secretary for Budget in accordance with Air Force Instruction (AFI) 65-503 are the prescribed cost factors to be used for "cost studies, economic analyses, component cost analyses, military construction projects, Program Objective Memorandum inputs, as well as programming, budgeting, accounting, and recording payments from other government agencies."[1]

Standard composite pay rates comprise the following elements: basic pay, RPA, basic allowance for housing (BAH), Medicare-Eligible Retiree Health Care (MERHC) accrual, basic allowance for subsistence (BAS), special and incentive pay, PCS, and miscellaneous pay.[2] Figure 2.1 shows how standard composite pay rates and their components vary by grade.[3] Basic pay accounts for the largest percentage of standard composite pay rates (45 percent), followed by BAH (17 percent), RPA (16 percent), miscellaneous (7 percent), MERHC (5 percent), BAS (4 percent), special and incentive pay (3 percent), and PCS (3 percent).

This breakdown suggests that workforce and personnel policy options that affect basic pay, BAH, and RPA have the greatest potential to reduce the average cost of an airman. This includes policies that directly affect pay and allowances, for example, limiting the rate of growth in compensation. This also includes policies that indirectly affect average pay and allowances, for example, shifting work to more-economical labor categories (e.g., by converting officer

---

[1] AFI 65-503, *Financial Management: US Air Force Cost and Planning Factors*, U.S. Department of the Air Force, July 13, 2018.

[2] As noted in AFI 65-503, Table A19-1, "Military Annual Standard Composite Pay," the standard composite pay rates "do not provide for the portion of military personnel benefits financed by other appropriations, such as the cost of government-furnished quarters for personnel residing in family housing or dormitories; the cost of mess attendant contracts for personnel subsisting in military dining facilities; and commissary and exchange benefits subsidized by appropriated funds."

[3] The average cost of an O-10 is slightly less than the cost of an O-9 because of the difference in BAH. Most O-10s live in base housing, which is excluded from standard composite pay rates.

requirements to enlisted), shifting to a more junior grade mix, or applying personnel policies to shift to a more junior years-of-service (YOS) mix.

Figure 2.1. Standard Composite Pay Rates by Grade (FY 2021)

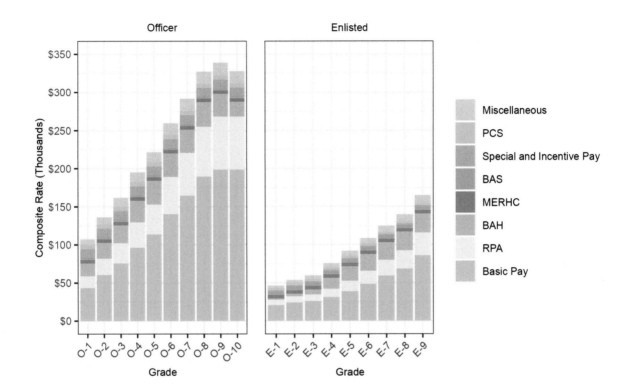

SOURCE: Adapted from AFI 65-503, 2018, Table A19-2, "Active Air Force Standard Composite Rates by Grade Description."

Historical Variations in MILPERS Spending

Air Force MILPERS spending and the average cost of an airman have varied over time. Between FY 2000 and FY 2021, Air Force spending on MILPERS roughly doubled, increasing from approximately $18 billion to $36 billion in then-year dollars (Figure 2.2, top). This reflects an average annual increase of 3.3 percent, outpacing annual growth in prices in the overall economy, which averaged 1.9 percent during the same period.[4]

---

[4] The FY 2000 active-duty MILPERS budget totaled $17,978,193,000; see DAF, 2001. Growth in prices in the overall economy between FY 2000 and FY 2021 is measured using the gross domestic product price deflator; see Office of the Under Secretary of Defense (Comptroller), 2021.

Figure 2.2. Air Force MILPERS Spending and End Strength over Time

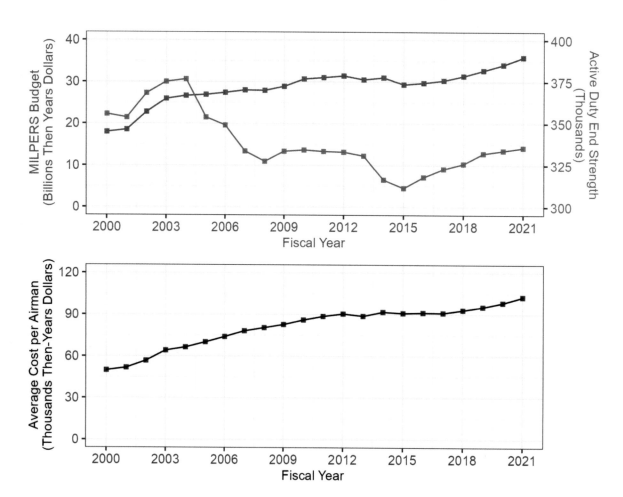

SOURCE: Adapted from DAF, "Military Personnel Program," *Air Force President's Budget FY22,* May 2021b.
NOTE: This figure represents the time span from FY 2000 to FY 2021.

The rate of growth is not attributable to changes in end strength. In fact, Air Force end strength *decreased* between FY 2000 and FY 2021 by approximately 6 percent, from 355,654 to 335,485 active-duty personnel (Figure 2.2, top).[5] The primary driver of growth in the Air Force MILPERS spending was the change in the average cost of an airman (Figure 2.2, bottom), which increased 3.5 percent annually in nominal terms, from approximately $50,000 to $103,000, between FY 2000 and FY 2021. By comparison, civilian pay increased by 2.3 percent annually during the same period.[6]

---

[5] The decline between FY 2000 and FY 2021 is less if one compares work years; work year declined from 360,226 in 2000 to 349,460 in 2021; see DAF, 2021b.

[6] Rates of civilian pay growth were calculated by RAND using Table 5.5 of Office of the Under Secretary of Defense (Comptroller), 2021.

Figure 2.2 illustrates how the average cost of an airman, MILPERS spending, and end strength relate to each other. As the cost of an airman increases, MILPERS purchasing power, or the number of personnel that a given dollar amount can afford, declines. If the rate of growth in the cost of an airman exceeds growth in MILPERS spending, end strength must also decline. The implication here is that workforce and personnel policies that decrease the cost of an airman may limit MILPERS spending without requiring end strength reductions. Absent such changes, the Air Force can limit MILPERS spending only by reducing end strength.

Conceptual Framework to Explore Policies to Limit Growth of the MILPERS Budget

MILPERS spending depends on the size (i.e., average strength) and makeup (e.g., grade strength, experience, and career field mix) of the personnel inventory. These factors are partially driven by demand, as expressed in funded authorizations.[7] They are also driven by policy (e.g., promotion timing) and external factors (e.g., the strength of the economy). The size and makeup of the actual inventory differs from funded authorizations because of constraints in the USAF human capital management system, along with external factors, such as the strength of the economy, that affect recruiting and retention. External factors may also shape the authorized workforce structure and personnel policies that the USAF selects.

The total cost of the MILPERS budget also depends on the average cost of an airman. The average cost of an airman partially reflects the makeup of the inventory. For example, standard composite pay rates are higher for officers than for enlisted personnel, and they increase with rank. As a result, workforce and personnel policies that affect the makeup of the personnel inventory also affect personnel costs. The average cost of an airman also reflects external factors like annual pay increases approved by Congress.[8]

The conceptual framework shown in Figure 2.3 represents the following dependencies:

- Authorized workforce structure—or the number of individuals authorized to serve by grade and Air Force Specialty Code (AFSC)—and personnel policies influence the makeup of the personnel inventory.
- The makeup of the personnel inventory influences the average cost of an airman.
- The size of the personnel inventory and the average cost of an airman influence MILPERS spending.

The framework is part of a larger dynamic system. For example, authorized workforce structure depends on inputs from functional areas, the Air Force's multilevel corporate structure, and Congress; the cost of an airman depends on annual changes to pay and allowances; and

---

[7] Albert A. Robbert, Lisa M. Harrington, Louis T. Mariano, Susan A. Resetar, David Schulker, John S. Crown, Paul Emslie, Sean Mann, and Gary Massey, *Air Force Manpower Determinants: Options for More-Responsive Processes*, RAND Corporation, RR-4420-AF, 2020.

[8] Lawrence Kapp, "Defense Primer: Military Pay Raise," Congressional Research Service, December 27, 2021.

personnel policies have additional costs outside the MILPERS budget (e.g., recruiting and training). Notwithstanding these simplifications, the framework provides a useful starting point for understanding the outcomes of different workforce and personnel policies.

Figure 2.3. Framework for Evaluating Options to Limit Growth of the MILPERS Budget

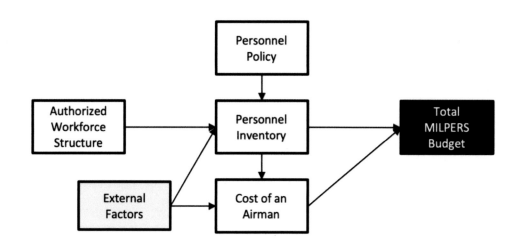

Limiting the Growth of the MILPERS Budget Is a Wicked Problem

Wicked problems are a class of planning and policy problems that are difficult to tackle because they lack clear definitions and boundaries, they involve complex interdependencies, and they do not have "correct" solutions.[9] Wicked problems cannot be solved in a purely analytic manner. They call for collaborative methods to enable information sharing, generate creative options, and build consensus around those options.

Table 2.1 lists the properties of wicked problems and links them to the challenge of identifying options to limit the growth of the MILPERS budget. For example, wicked problems lack a definitive formulation—the complete extent of personnel costs is ambiguous, and it is not clear whether the USAF should try to reduce the MILPERS appropriation, the direct cost of MILPERS, or the full cost of MILPERS.[10] Wicked problems do not have objectively "correct" solutions—because USAF stakeholders have different priorities in terms of managing manpower and personnel, operations, and logistics over different time horizons, they may prefer options that maximize different outcomes. Finally, wicked problems have consequential outcomes— workforce and personnel policy changes will affect service members in significant ways, and policymakers' actions may erode the USAF's ability to perform its missions.

---

[9] Horst W. J. Rittle and Melvin M. Webber, "Dilemmas in a General Theory of Planning," *Policy Sciences,* Vol. 4, No. 2, 1973.

[10] Seamus P. Daniels, *Assessing Trends in Military Personnel Costs*, Center for Strategic and International Studies, September 9, 2021.

Table 2.1. Properties of Wicked Problems Applied to MILPERS Spending

Property	Application to MILPERS Spending
There is no definitive formulation of the problem.	• Only about 80 percent of direct personnel costs are captured in the MILPERS budget. In addition, MILPERS does not capture indirect costs, such as variable O&M costs linked to workforce changes. Workforce and personnel policies may incur significant savings or expenses in other accounts. • Workforce and personnel policies may have significant nonfinancial effects (e.g., reduced personnel or training readiness) that limit the USAF's ability to accomplish its missions.
Solutions are not true or false, but good or bad.	• USAF stakeholders have different priorities and thus may judge policy options based on different outcomes and along different time horizons. • There is not an agreed-on level of savings in MILPERS spending that the USAF must achieve.
There is no immediate or ultimate test of the solution.	• The financial consequences of workforce and personnel policies unfold over many years and are not immediately obvious. • The nonfinancial consequences of workforce and personnel policies extend into areas that are harder to measure, such as training readiness.
The set of potential solutions is vast.	• Hundreds of factors directly and indirectly contribute to MILPERS spending, creating an effectively limitless solution space. • Factors may interact with one another, such that changes to one offset or amplify the effects of changes to another.
Every wicked problem is essentially unique.	• National defense objectives, USAF policies, the economy, and the military-eligible population change over time. These and other particulars may override similarities with historical attempts to influence MILPERS spending.
The planner has no right to be wrong.	• Workforce and personnel policies may have significant consequences for individuals in the USAF. • Workforce and personnel policies may also have significant national defense consequences.

As described in the introduction, standard approaches for making decisions about future workforce structure are data driven. These approaches use historic data, econometric models, staffing models, and cost and planning factors.[11] All these methods are useful for understanding and optimizing policies for *parts* of the overarching problem. However, because they necessarily entail simplification and abstraction, the answers they yield are imperfect. Furthermore, they stop short of making decisions, so the answers they yield are incomplete. In the next chapter, we describe wargaming as a complementary evidence-based approach to overcome some limitations of data-driven methods.

---

[11] AFI 65-503, 2018.

Chapter 3. Game Purpose and Structure

In this chapter, we describe the design of Operation Retrenchment Specter, a workforce futures policy game designed to explore options for reducing MILPERS costs.

Learning from Games

USAF policymaking should be evidence based.[1] Empirical data analysis is one example of an evidence-based approach used widely throughout the USAF. Empirical data analysis is important for MILPERS planning and programming. However, because MILPERS planning and programming is a wicked problem, empirical data analysis is unlikely to yield a complete solution.

Consider the example of an analyst examining how *high year tenure*, the maximum number of years that an enlisted service member may serve for at a given grade, may affect MILPERS costs. Using an inventory projection model, the analyst observes that reducing high year tenure produces a workforce with lower YOS on average, thereby reducing the average cost of an airman. However, stakeholders from the Air Force Recruiting Service object that the solution requires unobtainable recruiting goals, stakeholders from Air Education and Training Command object that accession training costs will wash out MILPERS savings, and stakeholders from the Office of the Deputy Chief of Staff for Operations (AF/A3) point out that the solution will result in unacceptable loss of experienced personnel. As this real-world example illustrates, empirical data analyses cannot capture all of the second-order effects, nor can they resolve which outcomes to optimize for.

Wargaming is another evidence-based approach used widely throughout the USAF. The U.S. Department of Defense (DoD) defines *wargaming* as the "representation of conflict or competition in a synthetic environment, in which people make decisions and respond to the consequences of those decisions."[2] More generally, national security policy analysis games involve players making decisions in a competitive environment with implicit and explicit rules, and then observing and dealing with the consequences of those decisions.[3]

We chose a game format to examine options to limit MILPERS costs for several reasons. First, games create an opportunity to generate new ideas and test novel options (i.e., the art of the

---

[1] DoD, *Executive Summary: DoD Data Strategy: Unleashing Data to Advance the National Defense Strategy*, September 30, 2020.

[2] Joint Chiefs of Staff, *Joint Planning*, Joint Publication 5-0, June 16, 2017, p. V-31.

[3] Elizabeth M. Bartels, *Building Better Games for National Security Policy Analysis: Towards a Social Scientific Approach*, RAND Corporation, RGSD-437, 2020.

possible).[4] Second, games are well suited to tackle problems that involve interactions between different parts of complex social systems (i.e., wicked problems), particularly when no single planner has complete knowledge of the system.[5] Third, games serve an educational purpose by allowing players from different backgrounds to share information and views. Fourth, games may transform players' perspectives, increasing consensus about issues and solutions.[6] For all these reasons, a workforce futures policy game is an ideal evidence-based approach for examining options to reduce MILPERS costs.

Even prior to conducting the game, the exercise of defining inputs, outcomes, and metrics with senior leaders and stakeholders from the Air Staff was tremendously informative. For example, because all potential policy levers could not be included in the simulation ecosystem, senior leaders first needed to articulate and prioritize their tacit views of which options were in play. This led to the inclusion of policy levers not initially considered, such as changes to compensation. As explained by one stakeholder, although the USAF does not control annual changes in basic pay, results from the game could be used to make a more informed argument to Congress for reforming how annual adjustments are calculated.[7]

In addition, because solutions could be evaluated along countless dimensions, senior leaders needed to articulate and prioritize their tacit views of what constituted a *good* solution. Once again, this led to the inclusion of outcomes not initially considered, such as the average cost of an airman. As noted by the same stakeholder, a solution that reduces total cost but not the average cost of an airman fails to increase MILPERS *buying power*, an important addendum. These scoping discussions also led to the inclusion of qualitative judgments of risk, as described later in this chapter.

Finally, the magnitudes of effects are difficult to comprehend. For example, a 1 percent change in cost, although small in relative terms, amounts to hundreds of millions of dollars. Scoping discussions with stakeholders revealed the importance of displaying absolute and relative changes in cost and anchoring dollar amounts to other operationally meaningful units of measure, such as flying hours.

Thus, an ancillary benefit of the approach is that *designing* the game forced clarity around how to make the underlying quantitative models and the simulation ecosystem more informative to decisionmakers.

---

[4] Bob Work and Paul Selva, "Revitalizing Wargaming Is Necessary to Be Prepared for Future Wars," *War on the Rocks*, December 8, 2015.

[5] Robert A. Levine, Thomas C. Schelling, and William M. Jones, *Crisis Games 27 Years Later: Plus C'est Deja Vu*, RAND Corporation, P-7719, 1991.

[6] Ed McGrady, "Getting the Story Right About Wargaming," *War on the Rocks,* November 8, 2019.

[7] Representatives from the Office of the Deputy Chief of Staff for Manpower, Personnel, and Services (AF/A1), interview with the authors, FY 2021. All interviewee quotes in this report stemmed from this material.

Game Design

In total, more than 50 individuals participated in Operation Retrenchment Specter. Individuals were divided among five teams (Gold, Black, Silver, Blue, and Green).[8] Each team consisted of a RAND moderator, a USAF leader, and a blend of USAF officers, enlisted personnel, and civilians drawn from different offices and backgrounds. This reflected a key objective of the game—to bring together individuals from diverse backgrounds to enable enterprise-wide thinking.[9]

The game was held in person over a single day and was divided into four rounds (Figure 3.1).

Figure 3.1. Structure of the Event

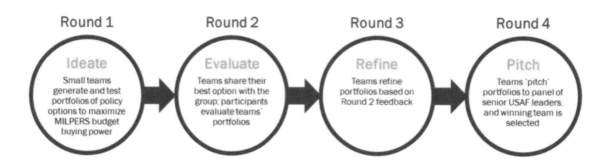

Round 1	Round 2	Round 3	Round 4
Ideate	Evaluate	Refine	Pitch
Small teams generate and test portfolios of policy options to maximize MILPERS budget buying power	Teams share their best option with the group; participants evaluate teams' portfolios	Teams refine portfolios based on Round 2 feedback	Teams 'pitch' portfolios to panel of senior USAF leaders, and winning team is selected

Round 1. Ideate

The objective of Round 1 was to generate portfolios of policy options to limit MILPERS costs. During the round, teams generated potential solution options. They entered these options into the simulation ecosystem—described in the next section—and reviewed the projected outcomes (i.e., the total cost, cost of an airman, full-time equivalents [FTEs], experience, and total accessions needed to sustain the workforce). Not all solutions could be entered into the simulation ecosystem. Nevertheless, teams were encouraged to discuss and capture these options for posterity.

After reviewing the outcomes, teams attempted to improve policy options with additional or adjusted proposed actions. Besides considering the outcomes, teams were instructed to think

---

[8] Six adjudicators came from AF/A1, AF/A3, AF/A4, AF/A5, AF/A8, and SAF/FM. Forty-five full-day participants came from AF/A1 (19), AF/A8 (3), AF/A5 (3), SAF/MR (2), AF/A4 (2), other DAF (9), and RAND (7). The distribution of grades among uniformed participants was one major general, one brigadier general, five colonels, nine lieutenant colonels, three majors, two captains, one lieutenant, and three chief master sergeants. The distribution among civilian participants was four senior executive service members, seven GS-15s and one GS-14.

[9] The diversity of participants, in terms of office, rank, and experience, was a strength of the exercise. However, future games could include additional participants, for example, career field managers, Major Command (MAJCOM) participants, and Air Reserve Component (ARC) participants.

about risks that the options entailed and to identify actions to mitigate those risks. At the conclusion of Round 1, each team developed a short pitch for their strongest solution option.

Teams were composed of officers, enlisted personnel, and civilians from different offices. Aside from the RAND moderator and the USAF leader, players were not assigned distinct roles. Moderators were given a general set of facilitation questions to probe potential solutions raised by the team: (1) How does the option relate to MILPERS spending? (2) What is the likely size of effect? and (3) What are the second-order effects? In addition, moderators were given schedules and resources to pace team discussions.

Round 2. Evaluate

The objective of Round 2 was to evaluate solution options. During the round, each team delivered an initial pitch to the full group. Pitches adhered to a prescribed form that included a brief description of the option; its strengths, weaknesses, and potential hedges against risks; changes to statutes or significant new policies that the option would require; and transition costs. To enable apples-to-apples comparison, teams were also required to display the top-line summary of the objective outcomes from the solution option automatically generated by the simulation ecosystem.

Following the initial pitches, participants evaluated all options along four dimensions:[10]

- *overall quality*—the overall ability of the option to limit MILPERS costs while avoiding unacceptable trade-offs, along with the *novelty* of the option
- *risk to mission*—the potential for the option to impede the USAF's ability to execute current, planned, and contingency operations over the next five years, as identified in the National Defense Strategy, and to hedge against shocks[11]
- *risk to force*—the potential for the option to impede the USAF's ability to recruit, maintain, train, equip, and sustain the workforce needed to meet strategic objectives
- *implementation risk*—the potential for outside factors, such as cost to implement; time to implement; and USAF, Office of the Secretary of Defense, or congressional resistance, to prevent the option from being adopted.

Each participant was given three up-votes to distribute among options they perceived as having the highest overall quality, and three down-votes to distribute among options they perceived as presenting the greatest risks.[12] To encourage novel options, up-votes were weighted by a factor of three (i.e., each up-vote was worth as much as three down-votes).

---

[10] *Risk to mission* and *risk to force* metrics are defined in Air Force Manual 90-1606, *Air Force Military Risk Assessment Framework*, U.S. Department of the Air Force, May 24, 2018. Although there are formal methods for assessing these forms of risk, we relied on professional military judgment because of the constraints of the policy game.

[11] A potential direction for future research is to evaluate options against different types, intensities, and durations of threats.

[12] During Round 1, teams were told that solutions would be evaluated along these four dimensions. During Round 2, participants were not allowed to vote for their own solution.

Voting occurred in an interactive fashion. Participants visited other teams and placed votes in four jars positioned on each team's table. Participants were encouraged to discuss their rationale and provide recommendations when casting down-votes.

At the conclusion of Round 2, the interim team standings were determined based on the difference between the total number of up-votes and down-votes for each proposed solution option.

Round 3. Refine

The objective of Round 3 was to refine solution options. During the round, teams debriefed feedback they received in Round 2 and revised their solution options. As in Round 1, they entered their solution options into the simulation ecosystem and reviewed the projected outcomes. Teams also identified additional hedges or safeguards to mitigate risk. At the conclusion of Round 3, each team developed a final pitch for their revised solution option.

Round 4. Pitch

The objective of Round 4 was to vet solution options. During the round, each team delivered a final pitch to a panel of senior USAF leaders, who served as adjudicators. Once again, participants and senior leaders evaluated options along four dimensions—overall quality, risk to mission, risk to force, and implementation risk—by casting up-votes and down-votes. Adjudicators' up-votes and down-votes were worth three times as much as participants' votes. At the conclusion of Round 4, the final team standings were determined.

Simulation Ecosystem

The workforce futures policy game was grounded in a simulation ecosystem that allowed teams to set workforce and personnel policies and observe the outcomes of those changes in real time. The simulation ecosystem combined elements of the conceptual framework shown in Figure 3.2, and it projected changes in the personnel inventory, MILPERS spending, and other financial and nonfinancial outcomes. The color gradients denote whether teams had complete, partial, or no control over the various elements. The simulation ecosystem borrowed components, or *modules*, from earlier work, which are described more fully in other reports.[13]

---

[13] See Matthew Walsh, David Schulker, Nelson Lim, Albert A. Robbert, Raymond E. Conley, John S. Crown, and Christopher E. Maerzluft, *Department of the Air Force Officer Talent Management Reforms: Implications for Career Field Health and Demographic Diversity*, RAND Corporation, RR-A556-1, 2021; and Matthew Walsh, Thomas Light, and Raymond E. Conley, *Assessing the Implications of Policy Options for the Military Personnel Budget: An Analytic Framework for Evaluating Costs and Trade-Offs*, RAND Corporation, RR-A1218-1, 2023.

Figure 3.2. Simulation Ecosystem and User-Enabled Inputs

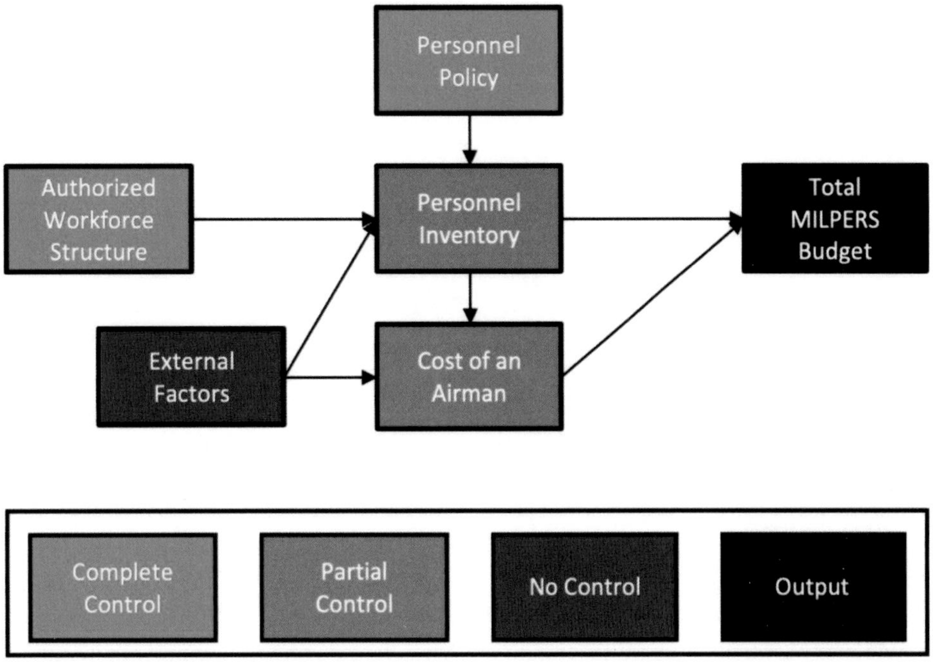

Simulation Ecosystem Inputs

Authorized Workforce Structure

The simulation ecosystem treated the authorized workforce structure as FY 2024–funded authorizations contained in the Manpower Programming and Execution System.[14] In addition to grade, each authorization has attributes related to occupation (e.g., AFSC), organization (e.g., unit, wing, group, and command), location (e.g., installation), command level (e.g., wing and above), and mission (e.g., tooth versus tail), among other things. The simulation ecosystem allowed teams to directly change the total number of authorizations and the mix of grades attached to these authorizations. The ecosystem also allowed teams to apply targeted changes to workforce segments based on attributes—for example, applying a 10 percent reduction to positions at the wing level and above. Of note, authorizations differ from the actual inventory. The simulation ecosystem allowed teams to apply changes to authorizations that, in conjunction with other factors, indirectly produced changes in the actual projected inventory.

---

[14] Funded authorizations include 63,376 officers and 268,158 enlisted personnel.

Personnel Policy and Inventory Projections

The simulation ecosystem contained structural models based on AFIs for officer and enlisted personnel promotion, separation, and retirement actions.[15] The ecosystem combined these with statistical models of promotion timing, promotion rates, and separation rates based on historical personnel data.[16] Teams used the ecosystem to simulate the long-run effects of workforce and personnel policies on the makeup of the personnel inventory. The ecosystem projected the number of officers and enlisted personnel by grade and YOS, and it estimated the number of annual accessions needed to sustain the workforce.

Average Cost of an Airman

The simulation ecosystem computed standard composite pay rates by grade and YOS. The ecosystem overlaid this table on the projected inventory to determine the average cost of an airman and total MILPERS spending. This allowed teams to observe how workforce and personnel policies affected MILPERS spending.

Year-to-year changes in basic pay are tied to the U.S. Department of Labor's Bureau of Labor Statistics Employment Cost Index (ECI).[17] Congress may, however, enact basic pay increases below the change in the ECI, as it has done in the past. The ecosystem allowed teams to increase or decrease military pay relative to civilian wages. These changes were linked to a retention elasticity factor such that retention moved in the direction of the change (e.g., increasing pay boosted retention).[18]

Training and Recruiting Costs

To approximate recruiting, accession, basic skills, and advanced training costs, we retrieved FY 2021 costs associated with the O&M Training and Recruiting budget activity ($930.5 million total).[19] In addition, we retrieved the number of officer and nonprior service enlisted gains in FY

---

[15] AFI 36-2032, *Military Recruiting and Accessions*, U.S. Department of the Air Force, September 27, 2019; AFI 36-2501, *Officer Promotion and Selective Continuation*, U.S. Department of the Air Force, April 30, 2021; AFI 36-2502, *Enlisted Airman Promotion and Demotion Programs*, U.S. Department of the Air Force, September 27, 2019; AFI 36-2606, *Reenlistment and Extension of Enlistment in the United States Air Force*, U.S. Department of the Air Force, September 20, 2019.

[16] Walsh et al., 2021.

[17] Kapp, 2021.

[18] John T. Warner, "The Effect of the Civilian Economy on Recruiting and Retention," *Report of the Eleventh Quadrennial Review of Military Compensation,* supporting research papers, Part 1, Chapter 2, U.S. Department of Defense, June 2012.

[19] Within the Accession Training budget subactivity, we retained SAG 031A (Officer Acquisition), 031B (Recruit Training), and 031D (Reserve Officers Training Corps). Within the Basic Skills and Advanced Training subactivity, we retained 032A (Specialized Skill Training). Finally, within the Recruiting and Other Education and Training subactivity, we retained 033A (Recruiting and Advertising), 033B (Examining), and 033E (Junior Reserve Officers Training Corps). These costs exclude flight training, reflecting an assumption that solutions that affect retention will not be applied, or will be applied along with some compensatory policy, to rated career fields.

2021 (32,510 total). We divided the total sum of recruiting and training costs by the total number of gains to calculate an index for the average cost of recruiting and training a new service member ($28,500 per accession).

The simulation ecosystem calculated the number of accessions needed to sustain the workforce and applied the recruiting and training index to estimate total annual accession costs.

Other Labor Categories

The simulation ecosystem allowed teams to shift work to other labor categories. Specifically, teams could increase the number of civilians, contractors, enlisted personnel, and officers in the ARC. The standard composite pay rates for these labor categories were $122,000, $200,000, $21,000, and $47,000, respectively.[20] Rates are significantly lower for ARC personnel because they contribute fewer labor days per year.

Simulation Ecosystem Outputs

Figure 3.3 shows summary outputs returned by the simulation ecosystem for a hypothetical solution option. The option in this case involved reducing military pay relative to civilian wages. The outputs are displayed for the user-defined inputs alongside a baseline that represents the force as closely as possible.

- *Total cost.* Separate values are shown for Regular Air Force (RegAF) officers and enlisted personnel, non-RegAF personnel (i.e., contractors, civilians, and ARC), and the cost to recruit and train new service members. Reducing compensation reduces the total cost of officers and enlisted personnel. The cost to recruit and train new service members increases, but the option still leads to a net decrease in total cost.
- *Cost of an airman.* Separate values are shown for officers and enlisted personnel. Reducing compensation reduces the average cost of officers and enlisted personnel.
- *FTEs.* Separate values are shown for officers, enlisted personnel, and non-RegAF personnel. Reducing basic pay does not affect FTEs because the number of annual accessions adjusts to sustain the workforce.
- *Experience (average YOS).* Separate values are shown for officers and enlisted personnel. Reducing compensation reduces retention, producing a more junior workforce with fewer average YOS.
- *Total accessions needed to sustain the workforce.* Separate values are shown for officers and enlisted personnel. Reducing compensation reduces retention, driving the need for more annual accessions to sustain the workforce.

---

[20] Brian T. Kelly, *Accelerate* Workforce *Change or Lose: An A1 Addendum to . . .* CSAF's Accelerate Change or Lose, white paper, Manpower, Personnel, and Services, U.S. Air Force, June 2021.

Figure 3.3. Quantitative Summary Provided by Analytic Tool for a Reduction in Basic Pay

	Solution	Baseline	Delta	Percent
Total Cost	$34,228,559,782	$35,023,243,541	($794,683,759)	(-2.3%)
Officer	$10,555,765,689	$10,732,728,958	($176,963,269)	(-1.6%)
Enlisted	$22,688,265,312	$23,401,820,525	($713,555,213)	(-3%)
Non-RegAF	$0	$0	$0	
Accessions	$984,528,781	$888,694,058	$95,834,724	(10.8%)
Cost of an Airman	$100,274	$102,960	($2,686)	(-2.6%)
Officer	$166,560	$169,353	($2,792)	(-1.6%)
Enlisted	$84,608	$87,269	($2,661)	(-3%)
Full Time Equivalent	331,533	331,533	0	(0%)
Officer	63,375	63,375	0	(0%)
Enlisted	268,158	268,158	0	(0%)
Non-RegAF	0	0	0	
Experience (YOS)	6.9	7.5	-0.5	(-7.1%)
Officer	8.5	8.8	-0.3	(-3.8%)
Enlisted	6.6	7.2	-0.6	(-8.1%)
Total Accessions	34,545	31,182	3,363	(10.8%)
Officer	4,652	4,352	299	(6.9%)
Enlisted	29,893	26,830	3,063	(11.4%)

NOTE: Screenshot from the simulation ecosystem. Cells in dark red and dark green denote a change from the baseline of greater than 5 percent. Cells in light red and light green denote a change from the baseline of 3 to 5 percent. Cells in peach denote a change from the baseline of less than 3 percent.

Game and Simulation Limitations

The results and findings from the game must be understood in light of assumptions, limitations, and design trade-offs inherent in this approach. The simulation ecosystem is grounded in empirical analysis of historic workforce data, yet the models it contains can only make (informed) predictions about future outcomes. In addition, because of the composition of players, the views expressed primarily (though not exclusively) are those of senior leaders working in the Air Staff or the Office of the Secretary of the Air Force. Moreover, all of the solutions offered come from the set supported by the simulation ecosystem, although teams were encouraged to discuss and record other options that could not be simulated. Finally, it was not feasible to develop detailed solutions in a single day. Thus, the outputs from the wargame reflect classes rather than point solutions.

Chapter 4. Proposed Solution Options

At the conclusion of Operation Retrenchment Specter, teams proposed a menu of options to senior USAF leaders. In this chapter, we describe those options. Chapter 5 includes further details about "the roads not taken."

For most solutions, teams were free to decide the size of the change to implement (e.g., what percentage of manpower to eliminate). Although this required professional military judgment, all teams included seasoned personnelists. Thus, although the solutions proposed would need to be further analyzed, they reflect plausible starting points.

Gold Team: Make Structural Modifications

The Gold team characterized the organizational structure of the Air Force as "plagued with bureaucracy." They saw developmental requirements as unduly influencing organizational structures. The team specifically called out lockstep requirements for professional military education and multiple stints in command.

With these inefficiencies in mind, the team proposed flattening Air Force organizational structures to create fewer echelons between operation and administration. They proposed a 10 percent reduction of the force above the wing level and a five percent reduction at and below the wing level. This would be achieved by consolidating MAJCOMs to reduce overhead in headquarters and combining units into fewer, larger wings. The team stressed that organizations do not need to be the same—rather, they can be tailored to suit the mission without unnecessary overhead. They maintained that by collapsing various missions and structures, the Air Force would be able to reduce manpower needs. The team advocated taking a deliberate, speedy approach to this reorganization so that savings could be realized as soon as possible.

The Gold team stressed that significant cost savings could be realized by reducing redundancies. An added benefit would be the reduction of tribalism that the team believed inhibits airmen growth and effectiveness. Airmen in broader, larger organizations would be exposed to multiple mission and functional areas and would therefore have an increased breadth of experience.

The team acknowledged that many of the options in their portfolio would be politically challenging to implement and that they would cut against an ingrained Air Force culture. There were potential risks to mission that the team felt might be difficult to quantify. They were also concerned that the portfolio ran the risk of repeating negative aspects of Program Budget

Decision 720 from the mid-2000s,[1] which reduced the force by 40,000 airmen for recapitalization and modernization. However, many felt that these cuts were arbitrary, that the promised efficiencies were not realized, and that missions were negatively affected.

Figure 4.1 shows the top-line summary for this solution option. Applying a 10 percent reduction to the workforce above the wing level and a five percent reduction at and below the wing level reduces annual overall spending by 6.3 percent ($2.2 billion). The reduction, as a percentage of baseline costs, is somewhat greater for officers because of the larger share making up positions above the wing level. The average cost of an airman also decreases slightly because of the more senior grade mix for positions above the wing level. The bulk of the savings, however, come from the 5.8 percent reduction in the number of FTEs. Finally, the total cost of accessions decreases because fewer airmen are needed to sustain the smaller workforce.

Figure 4.1. Analytic Outputs for the Gold Team Option

	Solution	Baseline	Delta	Percent
Total Cost	**$32,827,007,847**	**$35,023,243,541**	**($2,196,235,694)**	**(-6.3%)**
Officer	$9,972,162,490	$10,732,728,958	($760,566,468)	(-7.1%)
Enlisted	$22,016,855,808	$23,401,820,525	($1,384,964,718)	(-5.9%)
Non-RegAF	$0	$0	$0	
Accessions	$837,989,550	$888,694,058	($50,704,508)	(-5.7%)
Cost of an Airman	**$102,433**	**$102,960**	**($527)**	**(-0.5%)**
Officer	$169,142	$169,353	($211)	(-0.1%)
Enlisted	$86,908	$87,269	($361)	(-0.4%)
Full Time Equivalent	**312,292**	**331,533**	**-19,241**	**(-5.8%)**
Officer	58,957	63,375	-4,418	(-7%)
Enlisted	253,335	268,158	-14,823	(-5.5%)
Non-RegAF	0	0	0	
Experience (YOS)	**7.5**	**7.5**	**0.0**	**(-0.1%)**
Officer	8.8	8.8	0.0	(-0.2%)
Enlisted	7.2	7.2	0.0	(0%)
Total Accessions	**29,403**	**31,182**	**-1,779**	**(-5.7%)**
Officer	4,053	4,352	-299	(-6.9%)
Enlisted	25,350	26,830	-1,480	(-5.5%)

NOTE: Screenshot from the simulation ecosystem. Cells in dark red and dark green denote a change from the baseline of greater than 5 percent. Cells in light red and light green denote a change from the baseline of 3 to 5 percent. Cells in peach denote a change from the baseline of less than 3 percent.

Black Team: Blend, Broaden, Break, and Build

The Black team proposed consolidating AFSCs. They discussed *blending* (i.e., consolidating) AFSCs, which would allow for the broadening of skill sets of individual airmen to break down stovepipes in specialized communities, while building opportunities for enlisted airmen. The

---

[1] Doug Troyer, "Program Budget Decision 720, Force Shaping: Why Now??" Vance Air Force Base, August 28, 2007.

team estimated that consolidating AFSCs would enable a 2 percent manpower reduction. The team also considered targeted force reductions, including a 2 percent reduction in headquarters staffs. Finally, they proposed ending the payment of bonuses to members with over 12 YOS but not beyond 16 YOS.

The Black team maintained that the AFSC structure limits airmen's careers by forcing them to become too specialized, which prevents them from serving in other roles if desired or when needed. Additional opportunities for broadening skill sets were viewed as a net positive—airmen would feel more fulfilled in their jobs, improving overall satisfaction, and potentially improving retention.

The Black team proposed several potential actions to mitigate the risks associated with reducing end strength, for example, taking full advantage of developing multicapable airmen and using automation to increase workforce efficiencies. They also discussed leveraging other labor categories (active component, reserve components, and civilians) as a hedge, although they did not include the substitution of active component FTEs for reserve component or civilian FTEs in their final pitch. Finally, the team speculated that manpower reductions would force the identification and elimination of non–value-added work.

The Black team identified several implementation costs associated with their approach, for example, additional initial training for military members and upfront investment in the development, acquisition, and fielding of new technologies. They also observed that analysis would be needed to identify low-risk areas for manpower reduction. Finally, the team acknowledged that functional communities that have a vested interest in avoiding reductions would resist the option.

Figure 4.2 shows the top-line summary for this solution option. Reducing end strength by 2 percent reduces annual overall spending by 2 percent ($700.5 million). The savings come from the reduction in the number of FTEs along with the reduced number and total cost of accessions to sustain the smaller workforce. The average cost of an airman and experience are unchanged.

Figure 4.2. Analytic Outputs for the Black Team Option

	Solution	Baseline	Delta	Percent
Total Cost	$34,322,778,670	$35,023,243,541	($700,464,871)	(-2%)
Officer	$10,518,074,379	$10,732,728,958	($214,654,579)	(-2%)
Enlisted	$22,933,784,115	$23,401,820,525	($468,036,411)	(-2%)
Non-RegAF	$0	$0	$0	
Accessions	$870,920,177	$888,694,058	($17,773,881)	(-2%)
Cost of an Airman	$102,960	$102,960	$0	(0%)
Officer	$169,353	$169,353	$0	(0%)
Enlisted	$87,269	$87,269	$0	(0%)
Full Time Equivalent	324,902	331,533	-6,631	(-2%)
Officer	62,108	63,375	-1,267	(-2%)
Enlisted	262,795	268,158	-5,363	(-2%)
Non-RegAF	0	0	0	
Experience (YOS)	7.5	7.5	0.0	(0%)
Officer	8.8	8.8	0.0	(0%)
Enlisted	7.2	7.2	0.0	(0%)
Total Accessions	30,559	31,182	-624	(-2%)
Officer	4,265	4,352	-87	(-2%)
Enlisted	26,293	26,830	-537	(-2%)

NOTE: Screenshot from the simulation ecosystem. Cells in dark red and dark green denote a change from the baseline of greater than 5 percent. Cells in light red and light green denote a change from the baseline of 3 to 5 percent. Cells in peach denote a change from the baseline of less than 3 percent.

Silver Team: Pay the Right People and Be Ready for Crisis

The Silver team's solution option hinged on leveraging the high retention rates to increase the effectiveness of the force—in other words, a lean force can be effective if it is made up of the right people. The team proposed changing the incentive pay structure to shift to a more junior force. They proposed setting an incentive pay fund at 55 percent of the total annual pay increase. The fund would be used to incentivize members with mission-critical skills while avoiding the need to apply across-the-board pay increases. Under the team's solution option, enlisted airmen (E1 to E3) would continue to receive 100 percent of the annual increases, given concerns about their economic challenges.[2]

In addition to using incentives to retain people with mission-critical skills, the Silver team recommended applying other force management policies to shape the workforce skills mix. For example, the Air Force could use the Career Job Reservation program for first-term airmen, impose involuntary retraining under the Noncommissioned Officer Retraining Program, and apply high-year-of-tenure waivers in a targeted manner to ensure that the enlisted skills mix

---

[2] Mark Belinsky, "While Some Military Families Fight Hunger, Experts Propose Cutting Personnel Costs," *Military Officers Association of America*, December 1, 2021.

meets the Air Force's needs.[3] In conjunction with these changes, the Air Force could retain experience by facilitating the flow of active-duty members into the reserves through such programs as Palace Chase and Palace Front.[4]

The Silver team's option would result in a more junior—and less costly—officer and enlisted force. The approach would have other benefits besides reducing costs. For example, exercising more "teeth" in force management programs to keep the workforce young creates flexibility to grow the workforce in times of crisis by turning these measures off. Additionally, lowering retention in the RegAF while making it easier for individuals to transition into the reserves would increase the health of the ARC.

The team acknowledged that changes to compensation would require congressional approval and thus may be challenging to implement. In addition, monetary resources would be needed to establish the incentive pay fund. The team suggested implementing technology solutions to reduce manpower requirements and contributing the savings to the incentive pay fund. Finally, the team noted that further analysis would be necessary to establish the skill sets that the Air Force needs. The Air Force already targets special and incentive pay to certain workforce segments, such as pilots. However, identifying, tracking, and determining commensurate levels of compensation for more-granular skills and abilities would be a significant undertaking.

Figure 4.3 shows the top-line summary for this solution option. Limiting annual pay increases and shifting to a slightly more junior grade mix reduces annual overall spending by 1.3 percent ($450.3 million). The size of the workforce (i.e., FTEs) is unchanged: The savings come entirely from the 1.4 percent ($1,469) decrease in the average cost of an airman. Reducing pay reduces retention, which, along with shifting to a more junior grade mix, reduces average experience (YOS) by 2.9 percent (0.2 YOS). Finally, because individuals separate with fewer YOS on average, the number and total cost of accessions needed to sustain the workforce increase.

---

[3] All first-term airmen must have a Career Job Reservation to reenlist. The number of jobs available in each career field is set to prevent surpluses and shortages. To balance the enlisted career force across all AFSCs and ensure the sustainability of career fields, some noncommissioned officers are retrained (voluntarily or involuntarily) into other specialties each year. See AFI 36-2606.

[4] The Palace Chase program allows an active-duty member with an active-duty service commitment to serve a portion (usually one-third) of their remaining time in the reserve component. Palace Front is for members who are within 180 days of separation from the active component who want to transition to the reserve component without a break in service. See AFI 36-3205, *Applying for the Palace Chase and Palace Front Programs*, U.S. Department of the Air Force, September 10, 2008.

Figure 4.3. Analytic Outputs for the Silver Team Option

	Solution	Baseline	Delta	Percent
Total Cost	**$34,572,890,573**	**$35,023,243,541**	**($450,352,968)**	**(-1.3%)**
Officer	$10,633,655,837	$10,732,728,958	($99,073,121)	(-0.9%)
Enlisted	$23,013,869,821	$23,401,820,525	($387,950,705)	(-1.7%)
Non-RegAF	$0	$0	$0	
Accessions	$925,364,916	$888,694,058	$36,670,858	(4.1%)
Cost of an Airman	**$101,491**	**$102,960**	**($1,469)**	**(-1.4%)**
Officer	$167,789	$169,353	($1,563)	(-0.9%)
Enlisted	$85,822	$87,269	($1,447)	(-1.7%)
Full Time Equivalent	**331,533**	**331,533**	**0**	**(0%)**
Officer	63,375	63,375	0	(0%)
Enlisted	268,158	268,158	0	(0%)
Non-RegAF	0	0	0	
Experience (YOS)	**7.3**	**7.5**	**-0.2**	**(-2.9%)**
Officer	8.6	8.8	-0.2	(-2.1%)
Enlisted	6.9	7.2	-0.2	(-3.1%)
Total Accessions	**32,469**	**31,182**	**1,287**	**(4.1%)**
Officer	4,473	4,352	121	(2.8%)
Enlisted	27,996	26,830	1,166	(4.3%)

NOTE: Screenshot from the simulation ecosystem. Cells in dark red and dark green denote a change from the baseline of greater than 5 percent. Cells in light red and light green denote a change from the baseline of 3 to 5 percent. Cells in peach denote a change from the baseline of less than 3 percent.

Blue Team: Shrink the Overhead

The Blue team proposed reducing administrative overhead and restructuring parts of the Air Force to reduce overall MILPERS spending. For example, they challenged the practice of assigning five to six colonels to command positions at every wing. The team proposed shifting the overall grade distribution to the lower grades. This could be done in several ways: delaying promotion windows, adjusting retention bonuses, or reducing grade ceilings. In addition, the Blue team recommended consolidating MAJCOMs, eliminating Numbered Air Forces (NAFs), and shifting manpower to unfunded positions in field organizations.

As with the Silver team's solution option, the Blue team's solution option would result in a more junior, and less costly, workforce. A strength of this option is that the Air Force can change promotion timing and grade mix without congressional approval. However, organizational restructuring would require congressional approval and would drive significant transition costs. The Blue team acknowledged that these changes might reduce retention, creating recruiting challenges and introducing risk to mission because of its less experienced workforce.

Figure 4.4 shows the top-line summary for this solution option. Shifting to a more junior grade mix reduces annual overall spending by 1.6 percent ($558.8 million). The average cost of an airman decreases by 1.6 percent ($1,691), and average experience drops by 1.5 percent (0.1 YOS). Because individuals separate with fewer YOS on average, the number and total cost of annual accessions needed to sustain the workforce increase.

Figure 4.4. Analytic Outputs for the Blue Team Option

	Solution	Baseline	Delta	Percent
Total Cost	**$34,464,464,892**	**$35,023,243,541**	**($558,778,649)**	**(-1.6%)**
Officer	$10,268,900,823	$10,732,728,958	($463,828,135)	(-4.3%)
Enlisted	$23,305,056,159	$23,401,820,525	($96,764,367)	(-0.4%)
Non-RegAF	$0	$0	$0	
Accessions	$890,507,910	$888,694,058	$1,813,852	(0.2%)
Cost of an Airman	**$101,269**	**$102,960**	**($1,691)**	**(-1.6%)**
Officer	$162,034	$169,353	($7,319)	(-4.3%)
Enlisted	$86,908	$87,269	($361)	(-0.4%)
Full Time Equivalent	**331,533**	**331,533**	**0**	**(0%)**
Officer	63,375	63,375	0	(0%)
Enlisted	268,158	268,158	0	(0%)
Non-RegAF	0	0	0	
Experience (YOS)	**7.4**	**7.5**	**-0.1**	**(-1.5%)**
Officer	8.2	8.8	-0.6	(-6.5%)
Enlisted	7.2	7.2	0.0	(0%)
Total Accessions	**31,246**	**31,182**	**64**	**(0.2%)**
Officer	4,411	4,352	58	(1.3%)
Enlisted	26,835	26,830	6	(0%)

NOTE: Screenshot from the simulation ecosystem. Cells in dark red and dark green denote a change from the baseline of greater than 5 percent. Cells in light red and light green denote a change from the baseline of 3 to 5 percent. Cells in peach denote a change from the baseline of less than 3 percent.

Green Team: We Don't Need No Stinkin' Officers

The Green team's solution option focused on shifting force structure and duties from officers to enlisted personnel. They noted that the enlisted-to-officer ratio was roughly 4.2 to 1, and they proposed a 6-to-1 ratio instead. The team identified specialties in which a shift from officer to enlisted or officer to civilian was most feasible: remotely piloted aircraft pilot positions, nuclear and missile operations positions serving in missile silos, and support positions (e.g., force support, security forces, and civil engineering). The team did not recommend wholesale conversions of officer AFSCs to enlisted. Instead, they advocated for converting positions on a case-by-case basis.

The Green team justified their solution option by pointing out that other services have successfully undertaken similar conversions and that a growing number of enlisted personnel have completed undergraduate and advanced degrees. However, they recognized that the number of officer positions that could be converted to the enlisted force may be limited by the Uniform Code of Military Justice and that only nonmilitary essential positions could be civilianized. In addition, the team identified potential risk to mission in joint and coalition environments in terms of rank equivalency.

As an added cost savings initiative, the Green team recommended a monthly premium for health care coverage of $50 to be paid by every airman (or a sliding scale of payments equal to

$50 per member). However, the team admitted that it could be difficult to obtain congressional and sister service support for this added cost.

Figure 4.5 shows the top-line summary for this solution option. Increasing the officer-to-enlisted ratio to 6 to 1 reduces annual overall spending by 1.0 percent ($341.3 million). The average cost of an airman decreases by 1.0 percent ($1,042), reflecting the larger share of enlisted personnel making up the workforce. Total FTE is unchanged, though a portion of labor shifts from officers to enlisted personnel. Average experience decreases slightly, again because of the larger share of enlisted personnel, who tend to separate with fewer YOS. Finally, the number of accessions and the total cost to sustain the workforce increase slightly because of the larger share of enlisted personnel, who have slightly higher annual separation rates.

Figure 4.5. Analytic Outputs for the Green Team Option

	Solution	Baseline	Delta	Percent
Total Cost	$34,681,963,657	$35,023,243,541	($341,279,884)	(-1%)
Officer	$10,020,871,807	$10,732,728,958	($711,857,151)	(-6.6%)
Enlisted	$23,768,174,848	$23,401,820,525	$366,354,323	(1.6%)
Non-RegAF	$0	$0	$0	
Accessions	$892,917,002	$888,694,058	$4,222,944	(0.5%)
Cost of an Airman	$101,918	$102,960	($1,042)	(-1%)
Officer	$169,338	$169,353	($14)	(0%)
Enlisted	$87,269	$87,269	$0	(0%)
Full Time Equivalent	331,533	331,533	0	(0%)
Officer	59,177	63,375	-4,198	(-6.6%)
Enlisted	272,356	268,158	4,198	(1.6%)
Non-RegAF	0	0	0	
Experience (YOS)	7.5	7.5	0.0	(-0.3%)
Officer	8.8	8.8	0.0	(-0.2%)
Enlisted	7.2	7.2	0.0	(0%)
Total Accessions	31,330	31,182	148	(0.5%)
Officer	4,081	4,352	-272	(-6.2%)
Enlisted	27,250	26,830	420	(1.6%)

NOTE: Screenshot from the simulation ecosystem. Cells in dark red and dark green denote a change from the baseline of greater than 5 percent. Cells in light red and light green denote a change from the baseline of 3 to 5 percent. Cells in peach denote a change from the baseline of less than 3 percent.

Chapter 5. In-Game Evaluation of Solution Options

To summarize, the solution options generated by the five teams are:

- **Gold: make structural modifications.** Apply a 10 percent reduction to the workforce above the wing level and a five percent reduction at and below the wing level.
- **Black: blend, broaden, break, and build.** Consolidate AFSCs to enable a 2 percent reduction in end strength.
- **Silver: pay the right people.** Limit annual pay increases and shift to a more junior grade mix.
- **Blue: shrink the overhead.** Shift to a more junior grade mix.
- **Green: we don't need no stinkin' officers.** Convert officer positions to enlisted positions.

In this chapter, we compare the objective outcomes of these options. We then report participants' and adjudicators' subjective evaluations of the overall quality and risk of each. Finally, we present a thematic analysis of internal discussions that precipitated the selection of each team's solution option.

Objective Evaluations

Table 5.1 summarizes the outcomes of the solution options. The Gold and Black teams' options produce the largest annual savings, but they trade off size (i.e., FTE). The Silver and Blue teams' options produce the next largest savings, but they trade off experience (i.e., YOS). The Green team's option produces the smallest savings, but it does not trade off size or experience.

These options illustrate the rigid relationship between cost, size, and experience. Given the constraints of the dynamic system, all solutions involve trade-offs along one or more of these dimensions.

Table 5.1. Comparison of Outcomes from the Baseline Scenario and Potential Policy Options

Option	MILPERS Cost (billions of dollars)	Size (FTEs)	Experience (YOS)
Baseline	35.0	331,533	7.5
Gold: Flatten organizational structure to enable reductions in end strength.	32.8	312,292	7.5
Black: Consolidate AFSCs to enable reductions in end strength.	34.3	324,902	7.5
Silver: Limit annual pay increases and shift to a more junior grade mix.	34.6	331,533	7.3
Blue: Reduce administrative overhead to shift to a more junior grade mix.	34.5	331,533	7.4
Green: Convert officer positions to the enlisted force.	34.7	331,533	7.5

NOTE: Cells in dark red and dark green denote a 2.0 percent or more change from baseline. Cells in light red and light green denote a 1.0 percent or more change from baseline.

What do these savings amount to in practical terms? To give an example, the Blue team's solution option produces an annual savings of $558.8 million. Given that the average cost of an airman for this solution equals $101,269 and the average cost of a civilian equals $122,000, these savings could be repurposed to support 5,500 additional service members or 4,600 additional civilians. Alternatively, the average variable cost of F-16C and F-35A flying hours in FY 2021 equal $10,361 and $17,963, respectively.[1] The savings from the Blue team's solution option could be repurposed to support 53,900 additional F-16C flying hours, or 31,100 F-35A flying hours. Table 5.2 shows these conversions for each of the five solution options.

The key takeaway here is that although the savings from these options are modest in relative terms, they amount to significant values that could be converted, for example, into increased personnel or training readiness.

---

[1] Office of the Under Secretary of Defense (Comptroller), *Fiscal Year (FY) 2021 Department of Defense (DoD) Fixed Wing and Helicopter Reimbursement Rates,* October 1, 2020.

Table 5.2. Conversions of Savings from Potential Policy Options

Option	Savings (millions of dollars)	Extra Airmen (thousands)	Extra Civilians (thousands)	Extra F-16C Flying Hours (thousands)	Extra F-35A Flying Hours (thousands)
Gold: make structural modifications	2,196.2	n/a	18.0	212.0	122.3
Black: blend, broaden, break, and build	700.5	n/a	5.7	67.6	39.0
Silver: pay the right people	450.4	4.4	3.7	43.5	25.1
Blue: shrink the overhead	558.8	5.5	4.6	53.9	31.1
Green: we don't need no stinkin' officers	341.3	3.3	2.8	32.9	19.0

NOTE: Because the solutions proposed by the Gold and Black teams reduced the size of the workforce, we do not express the number of airmen that the savings could buy back. n/a = not applicable.

Subjective Evaluations

Table 5.3 shows scores based on the total numbers of votes that teams received from participants and adjudicators. The upper half of the table shows participant and adjudicator scores separately, and the lower half shows the combined scores. In terms of quality, the Silver and Gold teams had the highest scores (135 and 123, respectively). The Green team also received many votes from participants, whereas the Blue team received many votes from adjudicators.

In terms of risk, participants and adjudicators were primarily concerned with implementation. This does not imply that the solutions do not pose risk to force and mission but rather that participants and adjudicators perceived implementation risk as being more significant, or perhaps more obvious. The Silver and Black teams received the most votes for implementation risk. In the case of the Silver team, this was a result of the proposed reduction in compensation, which not only would be unpopular but also would require congressional approval. In the case of the Black team, this was a result of the proposed consolidation of AFSCs and the use of technology to increase workforce efficiencies, both of which have significant upfront costs and have had a mixed history of success. Participants perceived the Green team's proposal to convert officer to enlisted positions as posing the most-significant risk to force. Finally, participants and adjudicators perceived the Gold team's proposal to apply workforce reductions above, at, and below the wing level as posing the most-significant risk to mission. Overall, the Silver and Black teams had the lowest scores for risk (−42 and −39, respectively).

The Silver team received the highest score for quality and the lowest score for risk. As a result, the Gold team, which received the second highest score for quality, emerged as the overall leader (94), followed by the Silver team (93) and the Blue team (74).

Table 5.3. Summary of Subjective Evaluations

Source	Team	Quality[a]	Risk to Mission[b]	Risk to Force[b]	Implementation Risk[b]	Overall[c]
Participants/ Adjudicators	Gold	87/36	−4/−3	0/0	−19/−3	64/30
	Black	54/18	0/0	−2/0	−25/−12	27/6
	Silver	81/54	−1/0	−3/0	−20/−18	57/36
	Blue	54/45	−1/0	−7/−3	−8/−6	38/36
	Green	66/9	−1/0	−14/0	−9/−9	42/0
Combined Total	Gold	123	7	0	22	94
	Black	72	0	2	37	33
	Silver	135	1	3	38	93
	Blue	99	1	10	14	74
	Green	75	1	14	18	42

[a] Quality votes from participants and adjudicators were scaled by factors of 3 and 9, respectively.
[b] Risk votes from participants and adjudicators were scaled by factors of −1 and −3, respectively.
[c] Overall score reflects the quality score minus the sum of the three risk scores.

Participants and adjudicators from different organizations favored different options. Those from within AF/A1 favored the Silver team (45) followed by the Gold team (22), whereas those from outside AF/A1 favored the Gold team (72) followed by the Silver team (48). A key element of the Silver team's solution option was to increase special and incentive pay in a targeted manner to recruit and retain in-demand skills. This may have appealed to AF/A1 because they control special and incentive pay. The Gold team's solution option produced the largest monetary savings, which may have appealed to all individuals.

Thematic Analysis of Team Discussions

Over the course of the game, teams considered many options, only some of which appeared in their final pitches. To identify consistent findings from transcripts of team discussions, we used *thematic analysis*.[2] This is a method commonly used to identify, organize, and report patterns (i.e., themes) found within a large body of data in the social sciences. We coded relevant statements systematically across transcribed comments from team discussions. Table 5.4 summarizes all of the options discussed by the teams. Thematic analysis of their internal discussions provides additional insight into their thought processes.

---

[2] Jennifer Attride-Stirling, "Thematic Networks: An Analytic Tool for Qualitative Research," *Qualitative Research*, Vol. 1, No. 3, 2001, pp. 385–405.

Table 5.4. Options Discussed by Teams

Category	Option	Gold	Black	Silver	Blue	Green
Compensation	Slow growth in basic pay.	Not discussed	Not discussed	Included	Discussed	Discussed
Consolidation	Consolidate or eliminate MAJCOMs.	Included	Not discussed	Discussed	Included	Discussed
Consolidation	Consolidate or eliminate NAFs.	Included	Not discussed	Not discussed	Included	Discussed
Consolidation	Consolidate or eliminate wings.	Included	Not discussed	Not discussed	Not discussed	Not discussed
Consolidation	Consolidate AFSCs.	Included	Included	Discussed	Discussed	Not discussed
Consolidation	Divest excessive infrastructure (i.e., bases).	Not discussed	Discussed	Discussed	Discussed	Discussed
Consolidation	Reduce headquarters staff.	Included	Included	Not discussed	Not discussed	Not discussed
Health care	Charge health care premium.	Not discussed	Not discussed	Not discussed	Discussed	Included
PCS	Reduce frequency of PCSs.	Not discussed	Discussed	Discussed	Not discussed	Not discussed
Promotion	Move away from time-in-grade and time-in-service promotion system.	Discussed	Not discussed	Not discussed	Discussed	Not discussed
Promotion	Delay promotion timing.	Discussed	Included	Not discussed	Included	Discussed
Retention	Allow retention levels to decrease.	Discussed	Included	Included	Discussed	Not discussed
Retirement	Increase age of retirement.	Not discussed	Not discussed	Not discussed	Discussed	Discussed
Total Force	Shift to more junior grade mix.	Discussed	Discussed	Included	Included	Discussed
Total Force	Convert officer to enlisted.	Not discussed	Not discussed	Not discussed	Not discussed	Included
Total Force	Convert officers to warrant officers.	Not discussed	Discussed	Not discussed	Not discussed	Discussed
Total Force	Civilianize non-military-essential positions.	Not discussed	Not discussed	Discussed	Discussed	Discussed
Total Force	Reform ARC workforce structure.	Not discussed	Discussed	Discussed	Not discussed	Not discussed
Training	Shorten duration of training pipelines.	Not discussed	Discussed	Not discussed	Discussed	Discussed
Training	Reduce in-residence training and professional military education.	Not discussed	Not discussed	Discussed	Discussed	Discussed

NOTE: "Discussed" indicates that the option was discussed. "Included" indicates that the option was included in the final pitch. "Not discussed" indicates that the team did not consider this option.

Some teams discussed limiting growth in basic pay. As one participant shared, "Raises are great and our airmen deserve them, but other people in the economy aren't getting raises like this every year."[3] Teams proposed offsetting reductions in basic pay by increasing special and incentive pay. Doing so would allow the Air Force to incentivize in-demand skills. As another participant remarked, "Not every skill has to have the same pay scale." Teams acknowledged that this option may increase risk to force by hurting recruiting or retention and that it would require congressional approval.

All teams considered options to reduce manpower requirements by consolidating or eliminating organizations, installations, or functional communities. One participant questioned, "MAJCOMs, NAFs, and wings—do we really need them all?" Combining or eliminating MAJCOMs, NAFs, or wings would reduce manpower requirements in the costliest segments of the workforce—general and field grade officers. However, the number of individuals that a commander can effectively manage (i.e., span of control) might be exceeded for some positions. Divesting excess infrastructure would also reduce manpower requirements linked to base support functions. However, as one participant remarked, this option would present implementation risk because of the "political impact of closing down bases and of job loss." Finally, consolidating AFSCs could eliminate redundancies in career field management, in addition to creating economies of scale in training pipelines.[4] Several teams identified increased service member satisfaction and breadth of skills as secondary benefits of AFSC consolidation. One participant noted that a "less rigid classification system may allow airmen to develop additional skills in line with personal and career ambitions"; another said that "in addition to reducing costs, this may allow airmen to be more fulfilled in their job." Notwithstanding these benefits, teams pointed to the need for careful analysis to determine which AFSCs to consolidate and deliberate effort to ensure that airmen were not simply expected to do more. Teams also recognized that the up-front costs of consolidation would be significant.

Some teams considered delaying promotion timing. Raising the window of eligibility for promotion to field grade ranks could be used as part of an option to shift to a more junior grade mix. Doing so would reduce the rank associated with certain positions without reducing the experience of individuals assigned to those positions. One participant described it this way: "It's a cultural mindset that we 'need O6s'—by extending promotion, you get more experience." However, some teams questioned whether upskilling enlisted and junior personnel would make them more marketable in the civilian sector, potentially lowering retention. Some teams also discussed moving away from a promotion system based on time in grade and YOS. Although this would not directly affect MILPERS costs, it would have the secondary benefit of increasing retention of the most-talented airmen.

---

[3] Operation Retrenchment Spector policy game participants, interview with authors, December 13, 2021. All participant quotes in this report stem from this material.

[4] Raymond E. Conley and Albert A. Robbert, *Air Force Officer Specialty Structure: Reviewing the Fundamentals*, RAND Corporation, TR-637-AF, 2009.

No team set out to reduce retention, but many options (e.g., reduce compensation, delay promotion timing, and reduce or delay bonuses) had that effect. Teams recognized the cost of an experienced workforce, and they questioned whether reduced retention was necessarily bad. As one participant said, "We're close to the retention we achieved during stop loss"; another said, "We've built a retention force." Teams noted that in addition to reducing MILPERS costs, allowing retention to decrease in the RegAF may increase career field health in the ARC, and it would give the Air Force greater flexibility to expand the size of the workforce in times of crisis. Teams considered these benefits alongside the greater recruiting and training costs, and the risk to mission caused by loss of experience. However, loss of experience was not always viewed negatively. As one participant explained, "There are specialties that we might be interested in moving towards a younger force."

All teams considered ways to use more-economical labor categories and segments of the workforce. Shifting to a more junior grade mix would directly reduce MILPERS costs. Multiple teams asserted that the Air Force is underutilizing the talent and abilities of more junior service members. Converting officer positions to the enlisted force would also reduce MILPERS costs. Other services have successfully transitioned responsibilities from officers to enlisted personnel. Additionally, as one participant pointed out, "Take a look at the current enlisted force; they are more educated than ever before!"

Teams questioned which positions are truly military essential. One participant stated bluntly, "There are a lot of military personnel doing jobs that aren't necessarily military." Civilianizing certain officer positions could produce a net decrease across the military and civilian personnel budgets because of both the reduced cost of civilian personnel and the reduced number of days lost to training. However, teams acknowledged that this option was not applicable to all occupations and that it was economical for only officer positions. Finally, teams briefly touched on the options of creating a warrant officer corps or substantially reforming ARC force structure.

Most teams saw opportunities to shorten training pipelines by overhauling training curricula. As one participant frustratedly asked, "Take a look at the trainings—why are we teaching Excel?" Teams also saw opportunities to leverage technologies to accelerate training. One participant commented, "This is happening in [Air Education and Training Command] with immersive training pipeline—they've reduced the training pipeline by one-third by using augmented and virtual reality, and hands-on in the classroom." In addition, teams considered relaxing requirements for in-residence training and professional military education. All of these changes would reduce manpower requirements to support training and education, reduce PCS and temporary duty assignment costs, and decrease the annual number of student and transient personnel, yielding a manpower savings.[5]

Fewer teams considered changes to health care, noting that an annual premium, although small at the level of individual service members, would amount to significant savings across the

---

[5] These changes may yield additional savings to Air Education and Training Command.

force.[6] Some considered increasing the age of retirement. One participant asserted, "The airman of today does not think of a 20-year career; they think about doing a couple of years and then moving on." Increasing the age of retirement would reduce the number of individuals reaching retirement, as well as the number of years of retirement benefits that the Air Force would need to pay out to each individual. However, increasing the age of retirement has several drawbacks: It would require congressional approval; it may hurt recruiting and retention; and the resulting increase in the age of the workforce may increase risk to mission and medical costs.

Finally, some teams considered reducing PCS frequency, either by increasing assignment durations or by allowing remote work—an option that the USAF has recently begun to pursue. One participant mused, "I could live in Tampa, work remote for the Pentagon, and report to MacDill [Air Force Base]." This would reduce PCS costs, and it would also reduce the overall annual number of transient personnel, yielding a manpower savings.[7] Greater geographic stability may also increase service member satisfaction. However, increasing assignment durations may have negative retention effects at less desirable locations.

---

[6] The savings would likely be used to reduce the Defense Health Program O&M appropriation rather than the MILPERS budget.

[7] This would also produce savings in accounts used to pay for movement of household goods.

Chapter 6. Discussion and Recommendations

Spending on active-duty MILPERS has outpaced price growth in other areas of the economy since FY 2000. The size of the workforce decreased during this time. The primary driver of growth in the Air Force MILPERS spending, then, has been the change in the average cost of an airman. The growth in the average cost of an airman threatens to undermine the USAF's ability to field a ready workforce, while the resulting growth in the MILPERS budget threatens to divert resources from other components of readiness along with force modernization.

This is a wicked problem. Any action to limit MILPERS costs may have repercussive effects throughout the USAF enterprise. In addition, the problem lacks an objectively correct solution; USAF stakeholders have different priorities and thus may prefer options that optimize different outcomes and over different time horizons. Moreover, the consequences of actions are highly consequential at the levels of individual service members and national security. Because this is a wicked problem, it cannot be addressed using standard analytic methods, and it cannot be solved within the silo of manpower, personnel, and services.

To identify options to limit MILPERS costs, RAND formulated and conducted a workforce futures policy game—Operation Retrenchment Specter. The game was a success. Over 50 individuals from across the USAF participated. Options that the teams proposed were projected to yield annual savings ranging from $400 million to more than $2 billion. Teams considered risks associated with the options and developed hedges and mitigating courses of action. Although the options are still in the formative phase, they provide a basis for developing future actions to limit MILPERS costs.

Key Insights

Conclusion 1: There is a tight link between cost (i.e., MILPERS spending), size (i.e., FTE), and experience (i.e., YOS). Given that the size and composition of the inventory largely determines its cost, it is difficult to reduce MILPERS spending without trading off size or experience. This was evident across the set of options proposed by the teams:

- To trade off size in a potentially risk neutral manner, the Air Force could consider consolidating or eliminating organizations, installations, or functional communities. Doing so may remove redundancies without a loss of readiness.
- The Air Force could also create workforce savings to offset reduced size. One way to do this is to adopt technologies to increase workforce efficiencies. Another is to change policies or practices that reduce the number of days service members spend in student or transient status.
- To trade off experience in a potentially risk-neutral manner, the Air Force may more fully use the talent and abilities of airmen. This could be accomplished, for example, by

35

shifting to a more junior grade mix while increasing responsibilities of more junior service members or by converting officer positions to the enlisted force.

- Experience (i.e., YOS) is costly, both because of the structure of the military pay table and because of the bonuses needed to incentivize retention. In some career fields, the Air Force may be able to give up experience without increasing risk. Counterintuitively, fielding a less experienced force may also reduce a different type of risk: If retention rates were somewhat lower, the Air Force would have greater flexibility to rapidly increase the size of the force in times of crisis by applying measures (e.g., incentives or stop-loss) to restore very high retention levels.

Conclusion 2: Options considered by the teams entail significant implementation risk. Players overwhelming concentrated down-votes on implementation risk. Additionally, transcripts from team discussions captured skepticism arising from perceived implementation costs, cultural resistance, and decision authority:

- Implementation risk was seen as highest for solutions that require congressional approval, such as reducing pay increases or closing installations.
- However, the Air Force has the authority to immediately implement many of the options that were discussed, such as altering promotion timing; altering special and incentive pay; reducing grade ceilings; and, to some degree, consolidating organizational structures and functional communities. The Air Force has implemented several of these options but without the goal of reducing MILPERS costs. The impetus to reduce personnel costs has not yet been strong enough to overcome inertia to exercise these options for this purpose.

Conclusion 3: Several options can be significantly enhanced with hedging actions. From the second to the final round, teams succeeded in identifying several hedging actions to reduce risks associated with their solutions:

- For options that reduce retention in the RegAF, the Air Force could leverage programs to encourage additional service members to enter the ARC to retain their experience.
- For options that reduce basic pay, the Air Force could use special and incentive pay in a more targeted manner to sustain retention in high-demand career fields or in those with especially high production costs.
- For options involving shifting to a more junior grade mix, the Air Force could delay promotions to give individuals more time to gain experience.

Conclusion 4: Savings from small changes may be significant. The relative savings from most options were 2 percent or less. But because the MILPERS budget is so large, these are significant windfalls. A 2 percent savings could be repurposed to support thousands of additional civilians or service members or tens of thousands of flying hours.

Recommendations

Recommendation 1. Set up a "whole of Air Force" approach to limit MILPERS costs. Implementing options to limit MILPERS costs will have sweeping implications for the Air

Force. There are not only human capital, cultural, and quality-of-life implications but also force structure, infrastructure, readiness, programming and budgeting, and many other considerations. Although the AF/A1 is best positioned to lead efforts to explore options for limiting growth in MILPERS costs, authorizing and prioritizing the options that the Air Force will undertake must be decided on at the CSAF and Secretary of the Air Force level. There will be winners and losers; without clear lines of authority for decisionmaking, efforts will be stalled.

The Air Force should consider establishing a group that reports to CSAF that is dedicated to developing, vetting, and prioritizing actions to limit growth in the MILPERS budget. Because some potential options have long lead times, the group should implement options in a phased manner rather than waiting until a single all-encompassing plan is developed.

Recommendation 2. Further develop ideas proposed in the futures game. Given time constraints, teams effectively proposed solution classes rather than formal options.[1] For example, each organization and functional area is unique. How should MAJCOMs or NAFs be consolidated to limit risk and maximize manpower savings? Which functions or positions should be consolidated or converted to civilian? It may be time to loosen the constraints of standardized structures and objective organizations to allow for more-efficient arrangements. Accessing necessary data, demining cost-benefit trade-offs, and recommending appropriate courses of action will require dedicated analysis. Such effort is justifiable given the level of costs involved and the consequential nature of the course of action that may be adopted. One contribution of the game is the determination of solution classes that are promising enough to warrant such attention.

Recommendation 3. Continue to make use of policy games to explore this issue. Workforce futures policy games could be used as one method to perform the analysis recommended earlier. Policy games could also be used to develop additional innovative approaches and to encourage stakeholders to share information and views. Specific recommendations for next steps are:

- Develop specific scenarios, along with risk elements, and play them through with expert groups of participants over multiple days to obtain necessary data and to perform more-detailed analyses. For example, how do various solutions fare under alternate assumptions about future economic conditions that help or hinder recruiting and retention? How do they fare under alternate assumptions about the duration and intensity of contingencies? And how do they fare under alternate assumptions about budgetary pressures?
- Include players from different organizations and communities in future games to encourage new ideas and to establish more-widespread buy-in.

Recommendation 4. Add games to the set of evidence-based methods routinely used to examine workforce and personnel planning and policy issues. Games are already widely used

---

[1] For additional analysis of these and other options, see Walsh, Light, and Conley, 2023.

in the Air Force; however, they are not typically applied to workforce and personnel policy. This was one of the first examples of using a policy game in this way, and the results were complementary to those that have been used to generate other evidence-based approaches.

Controlling MILPERS spending is not the only wicked problem that Manpower, Personnel, and Services faces. On the contrary, most workforce and personnel planning problems are wicked. Policy games are a promising method to tackle these problems as well.

Finally, data-driven methods must continue to play a role in workforce and personnel planning. Analytic products generated using data-driven methods helped to ensure the game construct reflected real world trade-offs accurately. Additionally, as the Air Force narrowly considers some of the solution options, data-driven methods may once again become more applicable. Thus, we find that policy games may be most useful for directing attention to parts of the solution space, whereas data-driven methods may be most useful for optimizing options within those parts of the solution space.

Abbreviations

AFI	Air Force Instruction
AFSC	Air Force Specialty Code
ARC	Air Reserve Component
BAH	basic allowance for housing
CSAF	chief of staff of the Air Force
DAF	Department of the Air Force
DoD	U.S. Department of Defense
FTE	full-time equivalent
FY	fiscal year
MAJCOM	Major Command
MILPERS	military personnel
NAF	Numbered Air Force
O&M	Operation and Maintenance
PCS	permanent change of station
RegAF	Regular Air Force
RPA	retirement pay accrual
USAF	U.S. Air Force
YOS	year(s) of service

References

AFI—*See* Air Force Instruction.

Air Force Instruction 36-2032, *Military Recruiting and Accessions*, U.S. Department of the Air Force, September 27, 2019.

Air Force Instruction 36-2501, *Officer Promotion and Selective Continuation*, U.S. Department of the Air Force, April 30, 2021.

Air Force Instruction 36-2502, *Enlisted Airman Promotion and Demotion Programs*, U.S. Department of the Air Force, September 27, 2019.

Air Force Instruction 36-2606, *Reenlistment and Extension of Enlistment in the United States Air Force*, U.S. Department of the Air Force, September 20, 2019.

Air Force Instruction 36-3205, *Applying for the Palace Chase and Palace Front Programs*, U.S. Department of the Air Force, September 10, 2008.

Air Force Instruction 65-503, *Financial Management: US Air Force Cost and Planning Factors*, U.S. Department of the Air Force, July 13, 2018.

Air Force Manual 90-1606, *Air Force Military Risk Assessment Framework*, U.S. Department of the Air Force, May 24, 2018.

Attride-Stirling, Jennifer, "Thematic Networks: An Analytic Tool for Qualitative Research," *Qualitative Research*, Vol. 1, No. 3, 2001.

Bartels, Elizabeth M., *Building Better Games for National Security Policy Analysis: Towards a Social Scientific Approach*, RAND Corporation, RGSD-437, 2020. As of March 2, 2022: https://www.rand.org/pubs/rgs_dissertations/RGSD437.html

Bartels, Elizabeth M., Jeffrey A. Drezner, and Joel B. Predd, *Building a Broader Evidence Base for Defense Acquisition Policymaking*, RAND Corporation, RR-A202-1, 2020. As of August 4, 2022: https://www.rand.org/pubs/research_reports/RRA202-1.html

Belinsky, Mark, "While Some Military Families Fight Hunger, Experts Propose Cutting Personnel Costs," Military Officers Association of America, December 1, 2021.

Brown, Charles Q., Jr., *CSAF Action Orders to Accelerate Change Across the Air Force*, Chief of Staff, U.S. Air Force, February 7, 2022.

Conley, Raymond E., and Albert A. Robbert, *Air Force Officer Specialty Structure: Reviewing the Fundamentals*, RAND Corporation, TR-637-AF, 2009. As of March 2, 2022: https://www.rand.org/pubs/technical_reports/TR637.html

DAF—*See* U.S. Department of the Air Force.

Daniels, Seamus P., *Assessing Trends in Military Personnel Costs*, Center for Strategic and International Studies, September 9, 2021.

DoD—*See* U.S. Department of Defense.

Joint Chiefs of Staff, *Joint Planning*, Joint Publication 5-0, June 16, 2017.

Kapp, Lawrence, "Defense Primer: Military Pay Raise," Congressional Research Service, December 27, 2021.

Kelly, Brian T., *Accelerate* Workforce *Change or Lose: An A1 Addendum to . . .* CSAF's Accelerate Change or Lose, white paper, Manpower, Personnel, and Services, U.S. Air Force, June 2021.

Levine, Robert A., Thomas C. Schelling, and William M. Jones, *Crisis Games 27 Years Later: Plus C'est Deja Vu*, RAND Corporation, P-7719, 1991. As of March 2, 2022: https://www.rand.org/pubs/papers/P7719.html

McGrady, Ed, "Getting the Story Right About Wargaming," *War on the Rocks*, November 8, 2019.

Office of the Under Secretary of Defense (Comptroller), *Fiscal Year (FY) 2021 Department of Defense (DoD) Fixed Wing and Helicopter Reimbursement Rates*, October 1, 2020.

Office of the Under Secretary of Defense (Comptroller), *National Defense Budget Estimates for FY 2022*, (Green Book), August 2021.

Rittle, Horst W. J., and Melvin M. Webber, "Dilemmas in a General Theory of Planning," *Policy Sciences*, Vol. 4, No. 2, 1973.

Robbert, Albert A., Lisa M. Harrington, Louis T. Mariano, Susan A. Resetar, David Schulker, John S. Crown, Paul Emslie, Sean Mann, and Gary Massey, *Air Force Manpower Determinants: Options for More-Responsive Processes*, RAND Corporation, RR-4420-AF, 2020. As of January 20, 2022: https://www.rand.org/pubs/research_reports/RR4420.html

Troyer, Doug, "Program Budget Decision 720, Force Shaping: Why Now??" Vance Air Force Base, August 28, 2007.

U.S. Department of Defense, *Executive Summary: DoD Data Strategy: Unleashing Data to Advance the National Defense Strategy*, September 30, 2020.

U.S. Department of the Air Force, *FY 2002 Amended Budget Submissions to Congress June 2001: Operation and Maintenance, Air Force*, Vol. 1, June 2001.

U.S. Department of the Air Force, *Fiscal Year (FY) 2022 Budget Estimates: Military Personnel Appropriations*, May 2021a.

U.S. Department of the Air Force, "Military Personnel Program," *Air Force President's Budget FY22*, May 2021b. As of March 2, 2022:
https://www.saffm.hq.af.mil/FM-Resources/Budget/Air-Force-Presidents-Budget-FY22/

Walsh, Matthew, Thomas Light, and Raymond E. Conley, *Assessing the Implications of Policy Options for the Military Personnel Budget: An Analytic Framework for Evaluating Costs and Trade-Offs*, RAND Corporation, RR-A1218-1, 2023. As of March 31, 2023:
https://www.rand.org/pubs/research_reports/RRA1218-1.html

Walsh, Matthew, David Schulker, Nelson Lim, Albert A. Robbert, Raymond E. Conley, John S. Crown, and Christopher E. Maerzluft, *Department of the Air Force Officer Talent Management Reforms: Implications for Career Field Health and Demographic Diversity*, RAND Corporation, RR-A556-1, 2021. As of March 2, 2022:
https://www.rand.org/pubs/research_reports/RRA556-1.html

Warner, John T., "The Effect of the Civilian Economy on Recruiting and Retention," *Report of the Eleventh Quadrennial Review of Military Compensation*, supporting research papers, Part 1, Chapter 2, U.S. Department of Defense, June 2012.

Work, Bob, and Paul Selva, "Revitalizing Wargaming Is Necessary to Be Prepared for Future Wars," *War on the Rocks*, December 8, 2015.